SYSTEMS BEGINNERS

Everything you need to know about HVAC system

Kaylynn Lawson

Table of Contents

CHAPTER ONE 3

   INTRODUCTION 3

      Contents 4

   What does your HVAC system do? 5

      Your HVAC System's Parts ... 6

CHAPTER TWO 15

   CONDENSING UNIT AND EVAPORATOR COIL 15

      Ductwork 24

CHAPTER THREE 27

   KEEP YOUR HOME CLEAN 27

   Maintenance Fundamentals ... 29

CHAPTER ONE

INTRODUCTION

Are you aware of the modern marvel that is your HVAC system? An HVAC system that is properly maintained and functioning can provide comfort levels in your home that are beyond what generations before you could imagine. Modern HVAC systems can be controlled by your smartphone. They will keep you cool during hot summers and protect you from the coldest winter chills.

How much do you really know about the components and functions of your HVAC (Heating Ventilating and Air Conditioning) system? It is important to know the basics in order to make informed decisions when it comes time for you to purchase, upgrade, or maintain an HVAC system.

Contents

- What does your HVAC system do?
- Your HVAC System's Parts
- Furnace Types
- Keep Your Home Clean

- Maintenance Fundamentals

What does your HVAC system do?

Heating Ventilation Air Conditioning (HVAC) systems keep your home comfortable in winter and cool during summer. A central HVAC system controls all functions of the home. It is different than a system that heats and cools the home using a collection of heaters or window air conditioners. The central heating and cooling units circulate cool or warm air through a network of ducts to accomplish their mission.

Your HVAC System's Parts

There are many options available for HVAC systems. If you're thinking of installing or upgrading your HVAC system, there are many choices. These are the basic features you will find in most HVAC systems.

- **Furnace** - This unit heats your home's air. There are many energy sources that can be used to heat your air.

- **Heat exchanger** - This heat exchanger plays an important role heating the air in your home.

- **Condensing Units and Evaporator Coil** - These two components are the opposite of heat exchangers. They work together to cool your home's air.

- **Heat pump** - There are many heat pumps used in systems. They are an efficient way of heating and cooling your home and come in many different configurations.

- **Thermostat** - This crucial link in your system controls heating and cooling.

- **Ductwork** - The ductwork distributes warm or cold air throughout your home.

Let's get into the details of each component we have discussed.

Furnace Types

The furnace heats the air in an HVAC system. The furnace heats the air using an energy source. After that, the furnace's blower motor pushes air through the ductwork. Heating the air can be done with different energy sources. These are the top four energy sources.

Natural gas
The United States is a big fan of natural gas furnaces. More than half of American homes heat with natural gases. This number is even higher in some regions. The efficiency of natural gas furnaces is excellent, with newer models achieving over 90% efficiency. They can be extremely efficient depending on the price of natural gas. A natural gas furnace can only

be used if your home has access to natural gas lines.

Electric

Another common source of residential furnaces is electricity. Although electric furnaces may not be as efficient as gas furnaces in terms of efficiency, they are often cheaper to install and last a much longer time than gas furnaces. The cost of operating an electric furnace will vary depending on the cost of electricity in your area.

Propane

Propane is an appealing alternative to natural gas if you don't have any. Propane can be stored in tanks outside of your home, and delivery options are almost everywhere. Propane furnaces are used to heat about 10% of American homes.

Oil

Oil furnaces are still quite common in the northeastern United States. However, they are rare in other parts of the country. They are easy to install and efficient. To operate an oil furnace, oil must be delivered to your home and stored in a tank.

Heat Exchanger

Your furnace's heat exchanger is an essential component. Your furnace may look like a fireplace. An open flame heats the air and then it is sent to your home via ducts. It's actually a bit more complicated and safer thanks to the heat exchanger inside your furnace.

The furnace's gas flame and electric elements heat the air, but they are separate from the home's air. Heating takes place inside a sealed container. Air is circulated around the chamber to heat it. The heat exchanger is responsible for this circulation.

CHAPTER TWO

CONDENSING UNIT AND EVAPORATOR COIL

The heat exchanger heats the air moving through your home. However, the furnace components cool the air being circulated through your house.

The coils of an evaporator coil circulate chemical refrigerant fluids via a series. The refrigerants' expansion and contraction absorb heat from the air. The heat is then emitted from the refrigerants as they circulate through an outdoor condensing device. The refrigerants are then circulated back to the evaporator coil where the air is circulated. This will allow you to stay cool during hot summer months.

Heat Pumps

Heat pumps are a common feature of HVAC systems. Heat pumps are efficient and cost-effective as they use the temperature difference between your home and outside to heat or cool your home.

A heat pump circulates fluid between indoor and units using electricity. The fluid can either expand or contract to generate heat, or cool down. The heat pump transfers heat from outside to inside in winter. In summer, it reverses the process and moves heat from outside to inside. There are three types of heat pump systems.

- **Air source**
  This is the most popular

type of heat pump. This heat pump extracts heat from outside air to provide heating and cooling. They are both energy efficient and can be used to heat and cool. They are practical and economical in all climates.

- **Geothermal**

  Geothermal heat pump use the ground to heat their exchange medium. These systems are stable in subsurface temperatures, which makes them suitable for all seasons. They are therefore very reliable and highly efficient.

- **Water source**
  Water source heat pumps are quite efficient. These pumps move water through a continuous loop of water and use cooling towers to heat exchange. These systems are often used in large buildings such as office and apartment buildings, but are not suitable for single-family homes.

Heat pumps are not able to do the whole job in most cases. A heat pump-based HVAC system's indoor component will use auxiliary heating elements to increase heating in colder temperatures. Although most heating elements are electric, hybrid systems can sometimes use natural gas to provide additional heating.

Thermostat

Your thermostat is an essential part of your HVAC system. The thermostat regulates the temperature in your home. A thermostat is essential for controlling the temperature in your home. Without it, you would have to monitor it and manually turn on or off your cooling and heating systems. Modern thermostat technology has made it possible to choose from a variety of options when controlling your HVAC system.

- **These thermostats** can be programmed but are not programable. These thermostats allow you to set the desired temperature indoors and then leave it alone. They're great if you live at home and are comfortable making manual adjustments.

- **Programmable thermostats** offer automatic control options. The thermostat's controls allow you to set different temperatures for different times of day and even different days of week. You can also set the thermostat to lower the temperature at night if you prefer your home to be cooler when you go to sleep. You can also set the thermostat to adjust the temperature of your home to keep you away from it on certain days.

- **WiFi programmable thermostats** can be controlled from your smartphone or computer. This allows you to control the temperature of your home even when you aren't there.

- **Smart thermostats** allow you to schedule and remotely control your thermostat. They also track your patterns and behavior to optimize your heating or cooling.

A smart thermostat can make your home more efficient and help you save money.

Ductwork

After the air has been heated or cooled in your home, it needs to be circulated throughout. Your ducts or vents will do this. You'd find a number of pipes and metal boxes leading to and from your central air handler if you removed the walls of your house. These pipes supply the HVAC system with air, while others return heated or cooled air back to your home.

It is important to keep your home's ducts clean and free from leaks. This will ensure that you have optimal efficiency and good health. You should also ensure that the vents that bring air into your home are free from obstructions to maximize the benefits of your HVAC system. It's a good idea if your home is older to inspect the integrity and condition of your vents.

CHAPTER THREE

KEEP YOUR HOME CLEAN

Your HVAC system circulates the air in your home and cleans it. It is important to change the filters regularly to maintain clean air. However, there are ways to increase the air-cleaning abilities of your HVAC system.

- **Replace your filters.** You might consider using HEPA-rated or pleated filters. These filters can remove more dust, pollen and pet dander from the air than cheaper filters.

- **Install a central air purifier.** UV filters for the whole-home can kill bacteria and viruses, as well as reduce exposure to volatile organic compounds (VOC). These filters are a cost-effective way to make your home healthier.

Maintenance Fundamentals

It is easy to overlook your HVAC system until it becomes a problem. It's important to maintain your HVAC system in good condition. This will ensure that you are comfortable when it's cold and cool when it is hot. You'll reduce the need for emergency repairs and prolong the life of your HVAC system. These five tips can help you maintain your HVAC system.

- Check your filters regularly. Regularly changing your HVAC filters is an important part of maintaining clean air. Your equipment will last longer if you change the filters.
- Your condensing unit should be cleaned. Every season, clean the heat pump or condensing unit outside of your home. Use a regular water hose, not a pressure washer, to

clean the exterior of the unit.
- Make sure to inspect the drain pan and drain pipe of your evaporator coil. The drainpipe on your indoor units is for disposing of condensation. This drainpipe is essential to ensure that water doesn't back-up and cause damage to your home. To prevent unpleasant odors from forming, it's a good idea for your drain to be

sprayed with household bleach.
- Clear the area surrounding outdoor units. Keep your outdoor units free of debris and vegetation. Make sure you clean them regularly as part of your yard work.
- Professional maintenance is recommended. A HVAC system may appear to be working perfectly until it stops. Qualified HVAC technicians can perform

regular checks to identify and fix potential problems before they become serious.

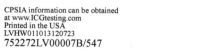